# Diary 2009

**Thraupis bonariensis**
Darwin collected this blue and yellow tanager
from Santa Fé while he was with HMS *Beagle*.

# More about Darwin ...

There are a huge number of online resources about Darwin and evolutionary theory, as well as places you might like to visit. Here is just a small selection.

### Darwin's life and work
Find out about Charles Darwin on *The Victorian Web*: *www.victorianweb.org*
Find out more about the life and times of Charles Darwin: *www.aboutdarwin.com*

### Darwin's writings online
Take a look at the most extensive collection of letters to and from Charles Darwin:
 *www.lib.cam.ac.uk/Departments/Darwin*
The Darwin-Wallace Paper as read out at the Linnean Society 1ˢᵗ July 1858: *www.linnean.org*
Read the complete works of Charles Darwin online: *www.darwin-online.org.uk*

### Darwin's heritage
Visit Darwin's birthplace, Shrewsbury: *www.darwinshrewsbury.org*
See Darwin's home, Down House, Luxted Road, Downe, Kent: *www.english-heritage.org.uk*

### Evolutionary theory
Understand the basics of evolution: *www.nhm.ac.uk/nature-online/evolution*
Explore more information on evolution: *www.evolution.berkeley.edu*
Find out more about the evolution controversy: *www.talkorigins.org*

### Key dates in Darwin's life
Born on the 12ᵗʰ February 1809
HMS *Beagle* departs from Plymouth 27ᵗʰ December 1831
HMS *Beagle* returns to Falmouth 2nd October 1836
Marriage to Emma 29ᵗʰ January 1839
*On the Origin of Species* published 1859
*The Descent of Man* published 1871
Died 19ᵗʰ April 1882

### Bicentenary events
Darwin exhibition on tour at the Natural History Museum, London
November 08 - March 09: *www.amnh.org/exhibitions/darwin*
Darwin200: *www.darwin200.org*
Darwin 2009 Festival, Cambridge, UK: *www.darwin2009.cam.ac.uk*
The Species of Origin contemporary Darwin Arts Project: *www.speciesoforigin.org*
The HMS *Beagle* Project: *www.thebeagleproject.com*
Shrewsbury Darwin Festival 2009: *www.darwinshrewsbury.org/about*

## JANUARY

| wk | M | T | W | Th | F | S | S |
|---|---|---|---|---|---|---|---|
| 1 |  |  |  | 1 | 2 | 3 | 4 |
| 2 | 5 | 6 | 7 | 8 | 9 | 10 | 11 |
| 3 | 12 | 13 | 14 | 15 | 16 | 17 | 18 |
| 4 | 19 | 20 | 21 | 22 | 23 | 24 | 25 |
| 5 | 26 | 27 | 28 | 29 | 30 | 31 |  |

## FEBRUARY

| wk | M | T | W | Th | F | S | S |
|---|---|---|---|---|---|---|---|
| 5 |  |  |  |  |  |  | 1 |
| 6 | 2 | 3 | 4 | 5 | 6 | 7 | 8 |
| 7 | 9 | 10 | 11 | 12 | 13 | 14 | 15 |
| 8 | 16 | 17 | 18 | 19 | 20 | 21 | 22 |
| 9 | 23 | 24 | 25 | 26 | 27 | 28 |  |

## MARCH

| wk | M | T | W | Th | F | S | S |
|---|---|---|---|---|---|---|---|
| 9 |  |  |  |  |  |  | 1 |
| 10 | 2 | 3 | 4 | 5 | 6 | 7 | 8 |
| 11 | 9 | 10 | 11 | 12 | 13 | 14 | 15 |
| 12 | 16 | 17 | 18 | 19 | 20 | 21 | 22 |
| 13 | 23 | 24 | 25 | 26 | 27 | 28 | 29 |
| 14 | 30 | 31 |  |  |  |  |  |

## APRIL

| wk | M | T | W | Th | F | S | S |
|---|---|---|---|---|---|---|---|
| 14 |  |  | 1 | 2 | 3 | 4 | 5 |
| 15 | 6 | 7 | 8 | 9 | 10 | 11 | 12 |
| 16 | 13 | 14 | 15 | 16 | 17 | 18 | 19 |
| 17 | 20 | 21 | 22 | 23 | 24 | 25 | 26 |
| 18 | 27 | 28 | 29 | 30 |  |  |  |

## MAY

| wk | M | T | W | Th | F | S | S |
|---|---|---|---|---|---|---|---|
| 18 |  |  |  |  | 1 | 2 | 3 |
| 19 | 4 | 5 | 6 | 7 | 8 | 9 | 10 |
| 20 | 11 | 12 | 13 | 14 | 15 | 16 | 17 |
| 21 | 18 | 19 | 20 | 21 | 22 | 23 | 24 |
| 22 | 25 | 26 | 27 | 28 | 29 | 30 | 31 |

## JUNE

| wk | M | T | W | Th | F | S | S |
|---|---|---|---|---|---|---|---|
| 23 | 1 | 2 | 3 | 4 | 5 | 6 | 7 |
| 24 | 8 | 9 | 10 | 11 | 12 | 13 | 14 |
| 25 | 15 | 16 | 17 | 18 | 19 | 20 | 21 |
| 26 | 22 | 23 | 24 | 25 | 26 | 27 | 28 |
| 27 | 29 | 30 |  |  |  |  |  |

## JULY

| wk | M | T | W | Th | F | S | S |
|---|---|---|---|---|---|---|---|
| 27 |  |  | 1 | 2 | 3 | 4 | 5 |
| 28 | 6 | 7 | 8 | 9 | 10 | 11 | 12 |
| 29 | 13 | 14 | 15 | 16 | 17 | 18 | 19 |
| 30 | 20 | 21 | 22 | 23 | 24 | 25 | 26 |
| 31 | 27 | 28 | 29 | 30 | 31 |  |  |

## AUGUST

| wk | M | T | W | Th | F | S | S |
|---|---|---|---|---|---|---|---|
| 31 |  |  |  |  |  | 1 | 2 |
| 32 | 3 | 4 | 5 | 6 | 7 | 8 | 9 |
| 33 | 10 | 11 | 12 | 13 | 14 | 15 | 16 |
| 34 | 17 | 18 | 19 | 20 | 21 | 22 | 23 |
| 35 | 24 | 25 | 26 | 27 | 28 | 29 | 30 |
| 36 | 31 |  |  |  |  |  |  |

## SEPTEMBER

| wk | M | T | W | Th | F | S | S |
|---|---|---|---|---|---|---|---|
| 36 |  | 1 | 2 | 3 | 4 | 5 | 6 |
| 37 | 7 | 8 | 9 | 10 | 11 | 12 | 13 |
| 38 | 14 | 15 | 16 | 17 | 18 | 19 | 20 |
| 39 | 21 | 22 | 23 | 24 | 25 | 26 | 27 |
| 40 | 28 | 29 | 30 |  |  |  |  |

## OCTOBER

| wk | M | T | W | Th | F | S | S |
|---|---|---|---|---|---|---|---|
| 40 |  |  |  | 1 | 2 | 3 | 4 |
| 41 | 5 | 6 | 7 | 8 | 9 | 10 | 11 |
| 42 | 12 | 13 | 14 | 15 | 16 | 17 | 18 |
| 43 | 19 | 20 | 21 | 22 | 23 | 24 | 25 |
| 44 | 26 | 27 | 28 | 29 | 30 | 31 |  |

## NOVEMBER

| wk | M | T | W | Th | F | S | S |
|---|---|---|---|---|---|---|---|
| 44 |  |  |  |  |  |  | 1 |
| 45 | 2 | 3 | 4 | 5 | 6 | 7 | 8 |
| 46 | 9 | 10 | 11 | 12 | 13 | 14 | 15 |
| 47 | 16 | 17 | 18 | 19 | 20 | 21 | 22 |
| 48 | 23 | 24 | 25 | 26 | 27 | 28 | 29 |
| 49 | 30 |  |  |  |  |  |  |

## DECEMBER

| wk | M | T | W | Th | F | S | S |
|---|---|---|---|---|---|---|---|
| 49 |  | 1 | 2 | 3 | 4 | 5 | 6 |
| 50 | 7 | 8 | 9 | 10 | 11 | 12 | 13 |
| 51 | 14 | 15 | 16 | 17 | 18 | 19 | 20 |
| 52 | 21 | 22 | 23 | 24 | 25 | 26 | 27 |
| 53 | 28 | 29 | 30 | 31 |  |  |  |

## JANUARY

| wk | M | T | W | Th | F | S | S |
|---|---|---|---|---|---|---|---|
| 1 |  |  |  |  | 1 | 2 | 3 |
| 2 | 4 | 5 | 6 | 7 | 8 | 9 | 10 |
| 3 | 11 | 12 | 13 | 14 | 15 | 16 | 17 |
| 4 | 18 | 19 | 20 | 21 | 22 | 23 | 24 |
| 5 | 25 | 26 | 27 | 28 | 29 | 30 | 31 |

## FEBRUARY

| wk | M | T | W | Th | F | S | S |
|---|---|---|---|---|---|---|---|
| 6 | 1 | 2 | 3 | 4 | 5 | 6 | 7 |
| 7 | 8 | 9 | 10 | 11 | 12 | 13 | 14 |
| 8 | 15 | 16 | 17 | 18 | 19 | 20 | 21 |
| 9 | 22 | 23 | 24 | 25 | 26 | 27 | 28 |

## MARCH

| wk | M | T | W | Th | F | S | S |
|---|---|---|---|---|---|---|---|
| 10 | 1 | 2 | 3 | 4 | 5 | 6 | 7 |
| 11 | 8 | 9 | 10 | 11 | 12 | 13 | 14 |
| 12 | 15 | 16 | 17 | 18 | 19 | 20 | 21 |
| 13 | 22 | 23 | 24 | 25 | 26 | 27 | 28 |
| 14 | 29 | 30 | 31 |  |  |  |  |

## APRIL

| wk | M | T | W | Th | F | S | S |
|---|---|---|---|---|---|---|---|
| 14 |  |  |  | 1 | 2 | 3 | 4 |
| 15 | 5 | 6 | 7 | 8 | 9 | 10 | 11 |
| 16 | 12 | 13 | 14 | 15 | 16 | 17 | 18 |
| 17 | 19 | 20 | 21 | 22 | 23 | 24 | 25 |
| 18 | 26 | 27 | 28 | 29 | 30 |  |  |

## MAY

| wk | M | T | W | Th | F | S | S |
|---|---|---|---|---|---|---|---|
| 18 |  |  |  |  |  | 1 | 2 |
| 19 | 3 | 4 | 5 | 6 | 7 | 8 | 9 |
| 20 | 10 | 11 | 12 | 13 | 14 | 15 | 16 |
| 21 | 17 | 18 | 19 | 20 | 21 | 22 | 23 |
| 22 | 24 | 25 | 26 | 27 | 28 | 29 | 30 |
| 23 | 31 |  |  |  |  |  |  |

## JUNE

| wk | M | T | W | Th | F | S | S |
|---|---|---|---|---|---|---|---|
| 23 |  | 1 | 2 | 3 | 4 | 5 | 6 |
| 24 | 7 | 8 | 9 | 10 | 11 | 12 | 13 |
| 25 | 14 | 15 | 16 | 17 | 18 | 19 | 20 |
| 26 | 21 | 22 | 23 | 24 | 25 | 26 | 27 |
| 27 | 28 | 29 | 30 |  |  |  |  |

## JULY

| wk | M | T | W | Th | F | S | S |
|---|---|---|---|---|---|---|---|
| 27 |  |  |  | 1 | 2 | 3 | 4 |
| 28 | 5 | 6 | 7 | 8 | 9 | 10 | 11 |
| 29 | 12 | 13 | 14 | 15 | 16 | 17 | 18 |
| 30 | 19 | 20 | 21 | 22 | 23 | 24 | 25 |
| 31 | 26 | 27 | 28 | 29 | 30 | 31 |  |

## AUGUST

| wk | M | T | W | Th | F | S | S |
|---|---|---|---|---|---|---|---|
| 31 |  |  |  |  |  |  | 1 |
| 32 | 2 | 3 | 4 | 5 | 6 | 7 | 8 |
| 33 | 9 | 10 | 11 | 12 | 13 | 14 | 15 |
| 34 | 16 | 17 | 18 | 19 | 20 | 21 | 22 |
| 35 | 23 | 24 | 25 | 26 | 27 | 28 | 29 |
| 36 | 30 | 31 |  |  |  |  |  |

## SEPTEMBER

| wk | M | T | W | Th | F | S | S |
|---|---|---|---|---|---|---|---|
| 36 |  | 1 | 2 | 3 | 4 | 5 |  |
| 37 | 6 | 7 | 8 | 9 | 10 | 11 | 12 |
| 38 | 13 | 14 | 15 | 16 | 17 | 18 | 19 |
| 39 | 20 | 21 | 22 | 23 | 24 | 25 | 26 |
| 40 | 27 | 28 | 29 | 30 |  |  |  |

## OCTOBER

| wk | M | T | W | Th | F | S | S |
|---|---|---|---|---|---|---|---|
| 40 |  |  |  |  | 1 | 2 | 3 |
| 41 | 4 | 5 | 6 | 7 | 8 | 9 | 10 |
| 42 | 11 | 12 | 13 | 14 | 15 | 16 | 17 |
| 43 | 18 | 19 | 20 | 21 | 22 | 23 | 24 |
| 44 | 25 | 26 | 27 | 28 | 29 | 30 | 31 |

## NOVEMBER

| wk | M | T | W | Th | F | S | S |
|---|---|---|---|---|---|---|---|
| 45 | 1 | 2 | 3 | 4 | 5 | 6 | 7 |
| 46 | 8 | 9 | 10 | 11 | 12 | 13 | 14 |
| 47 | 15 | 16 | 17 | 18 | 19 | 20 | 21 |
| 48 | 22 | 23 | 24 | 25 | 26 | 27 | 28 |
| 49 | 29 | 30 |  |  |  |  |  |

## DECEMBER

| wk | M | T | W | Th | F | S | S |
|---|---|---|---|---|---|---|---|
| 49 |  |  | 1 | 2 | 3 | 4 | 5 |
| 50 | 6 | 7 | 8 | 9 | 10 | 11 | 12 |
| 51 | 13 | 14 | 15 | 16 | 17 | 18 | 19 |
| 52 | 20 | 21 | 22 | 23 | 24 | 25 | 26 |
| 53 | 27 | 28 | 29 | 30 | 31 |  |  |

29 Monday

30 Tuesday

31 Wednesday                                                            New Year's

I  Thursday                                                             New Year's

2  Friday

3  Saturday

4  Sunday

1837  The Zoological Society of London
receives 80 mammals and 450 birds,
including the finches, collected by Darwin
during the voyage of HMS *Beagle*.

**Darwin specimens**
Thousands of specimens were collected by Darwin on his
voyage and brought back to England for further study. Some
of the zoological specimens are now stored in spirit jars in
the Darwin Centre of the Natural History Museum, London

5   Monday

---

6   Tuesday

Epiph

---

7   Wednesday

1860 The second edition of *On the Origins of Species* is published and 3000 copies printed.

---

8   Thursday

---

9   Friday

---

IO   Saturday

Islamic New Y

---

II   Sunday

Full Moo

---

**The works of Charles Darwin**
A selection of various editions of Darwin's writings. Darwin wrote sixteen books and many editions of some of the titles. All his writings remain in publication today in various forms and languages.

ORIGIN OF SPECIES

ORIGIN OF SPECIES DARWIN

DARWIN THE ORIGIN OF SPECIES

ISBN 0 14
040.001 X

Charles Darwin • The Origin of Species

09212

The
Origin
of
Specie

Darwin

ORIGIN
OF
SPECIES DARWIN

A MONOGRAPH
OF THE
CIRRIPEDIA
BALANIDÆ

DARWIN

1854

## I2  Monday

1836  HMS *Beagle* arrives in Sydney, Australia.

## I3  Tuesday

## I4  Wednesday

## I5  Thursday

## I6  Friday

1836  Darwin sets off for Bathurst, 100 miles
west of Sydney. He saw the lack of kangaroos
as evidence of how damaging European
settlement had been to the wildlife.

## I7  Saturday

## I8  Sunday

**HMS *Beagle***
The upper deck, and fore and aft of the middle section of
HMS *Beagle*, a ten gun, 30 m (100 ft) long brig. Life on
board was extremely cramped; Darwin shared a poop cabin
that measured only 3 m² (11 ft²) with two other men.

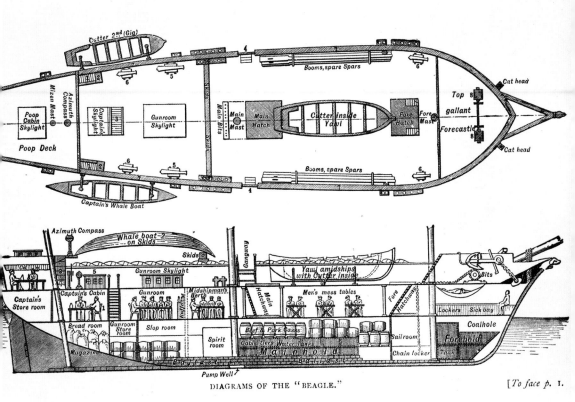

Cutter 2nd (Gig)

Poop Cabin Skylight

Poop Deck

Mizen Mast

Azimuth Compass

Captain's Skylight

3

Gunroom Skylight

Main Bits

Skids

Main Mast

Main Hatch

Main Hatch

4

Booms, spare Spars

Cutter inside Yawl

Fore Hatch

Fore Mast

Top gallant Forecastle

8

8

Cat head

Cat head

6

5

6

6

5

6

Captain's Whale Boat

4

Booms, spare Spars

Azimuth Compass

Whale boat-2 on Skids

Skids

Ganrway

Yawl amidships with Cutter inside

Bits

3

5

Gunroom Skylight

Captain's Store room

Captain's Cabin

Bread room

Gunroom Store room

Gunroom

Slop room

Midshipman's Berth

Spirit room

Beef & Pork Casks

Cable tiers Water-Tanks

Mainhold

Main Hatchway

Men's mess tables

Fore Hatchway

Lockers

Sick bay

Coalhole

Sailroom

Chain locker

Tank

Forehold

Magazine

Pump Well

DIAGRAMS OF THE "BEAGLE."

[To face p. 1.

**19** Monday

Martin Luther King, Jr. Day (U

---

**20** Tuesday

---

**21** Wednesday

---

**22** Thursday

---

**23** Friday

---

**24** Saturday

**1839** Darwin is elected as a Fellow of the Royal Society of London.

---

**25** Sunday

---

***Paradisea rubra***
Alfred Russel Wallace (1823-1913) studied the red bird of paradise whilst developing his own theory of evolution through natural selection. He noted it was "the only species that inhabit Waigiou and is peculiar to that island". Like Darwin, Wallace wa observing how isolation could give rise to new species.

## 26 Monday

Lunar New Year, New Moon

## 27 Tuesday

## 28 Wednesday

## 29 Thursday

1839 Emma Wedgwood and Charles
Darwin are married at St. Peter's
Anglican Church in Maer, Staffordshire.

## 30 Friday

## 31 Saturday

## I Sunday

**Annie's box**
After her beloved daughter's death, Emma Darwin filled
Annie's writing box with some of her childhood possessions
together with a map, showing the location of Annie's grave
in Malvern, and notes by Charles on his daughter's illness.

# February 2009

2   Monday

3   Tuesday

4   Wednesday

5   Thursday

6   Friday

1796  Birth of John Stevens Henslow,
Darwin's mentor at Cambridge and
lifelong friend. Darwin became known
as "the man who walks with Henslow".

7   Saturday

8   Sunday

**John Stevens Henslow, 1796-1861**
Probably the most influential figure on Darwin in the years
before his voyage was John Henslow. He tutored Darwin
in natural history at Cambridge and recommended him as
a suitable companion for Captain FitzRoy on HMS *Beagle*.

J. S. Henslow.

9   Monday

10   Tuesday

1869   The fifth edition of *On the Origin of Species* is published. For the first time the phrase coined by Herbert Spencer, "survival of the fittest", is included.

11   Wednesday

12   Thursday

1809   Birth of Charles Robert Darwin at The Mount, Shrewsbury, Shropshire.

13   Friday

14   Saturday

15   Sunday

**Charles Robert Darwin, 1809-1882**
An illustrated portrait of Darwin, *c.*1840s.

# February 2009

**16** Monday

President's Day (

**17** Tuesday

**18** Wednesday

**19** Thursday

1872 The sixth edition of *Origin of Species* is published. For the first time, a glossary and the word "evolution" are included, and "On" is dropped from the title.

**20** Friday

**21** Saturday

**22** Sunday

**Archaeopteryx lithographica**
The fossil specimen of *Archaeopteryx*, with its feathered wings was an important discovery in 1862, providing evidence of a between dinosaurs and birds. It is mentioned in the fourth edition of *On the Origin of Species* as an example of a transiti fossil that helped confirm Darwin's theory of common desce

23 Monday

24 Tuesday                                                      Shrove Tuesday (Christi

25 Wednesday                                        Ash Wednesday (Christian), New Moor

26 Thursday

27 Friday

28 Saturday

1832 Darwin sets foot for the first time
on the continent of South America, his
mind a "chaos of delight" with the
beauty of nature.

I Sunday                                                        St. David's Day (Wal

**The Descent of Man, New York, 1923**
Darwin published this work in 1871 and during his lifetime it
was translated into Danish, Dutch, French, German, Italian,
Polish, Russian and Swedish. Since his death, there have bee
many other translations including a Yiddish edition in 1923.

2   Monday

3   Tuesday

4   Wednesday

5   Thursday

6   Friday

7   Saturday

1837 Darwin leaves Cambridge and takes
lodgings in Great Marlborough Street, London.
Here, "...I gradually came to disbelieve in
Christianity as a divine revelation".

8   Sunday

Daylight Saving Time begins (UK & U

**Charles Darwin in South America**
A modern depiction of Darwin with the *Beagle* in the
background, for a 1979 edition of *The Journal of a Voyage
in HMS Beagle* by Charles Darwin.

9   Monday
1869 Alfred Russel Wallace publishes
*The Malay Archipelago* chronicling his
travels and scientific exploration there.
He dedicates it to Charles Darwin.

10   Tuesday

11   Wednesday

Full Moon

12   Thursday

13   Friday

14   Saturday

15   Sunday

**Orang utan attacked by Dyaks from *The Malay Archipelago***
Alfred Russel Wallace formed his ideas on evolution at much
the same time as Darwin. Today, Wallace is remembered for
being the father of zoogeography rather than for the concept
of natural selection. Both theories were a product of his eight
years study in Malaysia and Indonesia.

16 Monday

17 Tuesday                                                   St. Patrick's D

18 Wednesday

19 Thursday

20 Friday                                     Vernal Equinox – Spring begins (UK & U

**March 1835** Climbing the Andes at Valparaíso, Darwin finds petrified trees similar to those at sea level. He becomes certain the mountains rose "slowly and by little starts".

21 Saturday

22 Sunday                                                   Mother's Day (L

**HMS *Beagle***
Darwin was on board HMS *Beagle* as a "gentleman's companion" to the commander Captain FitzRoy, who feared the loneliness of his voyage to chart the coast of South America might drive him mad.

H.M.S. BEAGLE IN STRAITS OF MAGELLAN.   MT. SARMIENTO IN THE DISTANCE.       *Frontispiece.*

## 23 Monday

## 24 Tuesday

## 25 Wednesday

**March 1861** The third edition of *On the Origin of Species* is published and includes, for the first time, a historical introduction listing the precursors of evolutionary thought.

## 26 Thursday

New Moon

## 27 Friday

## 28 Saturday

## 29 Sunday

**Pseudoscarus lepidus**
This specimen of a parrot fish was collected by Darwin when he was in Tahiti and is one of many specimens brought back from the voyage of HMS *Beagle*.

*Pseudoscarus lepidus*

Tahiti

30 Monday

31 Tuesday

1 Wednesday

**Spring 1855** Darwin builds a bird house in his garden and starts breeding pigeons to help with his study on selection.

2 Thursday

3 Friday

4 Saturday

5 Sunday

Palm Sunday (Christia

**Fancy pigeons**
In 1855 Darwin began a serious study of domestic animals, including pigeons. He visited the beer-halls and local haunts of pigeon-fanciers to learn about artificial selection and bega breeding pigeons himself.

## 6 Monday

**April 1836** Darwin studies coral atolls on
the Cocos Islands to test his theory of reef
formation; "...such formations surely rank high
amongst the wonderful objects of this world."

## 7 Tuesday

## 8 Wednesday

Passover (Jewis

## 9 Thursday

Maundy Thursday (Christian), Full Moon

## 10 Friday

Good Friday (Christia

## 11 Saturday

**Between February and April 1826**
While at Edinburgh University, Darwin
takes lessons in taxidermy from John
Edmonstone, a freed slave.

## 12 Sunday

Easter Sunday (Christia

**Beak variations in finches**
Such variations within the same group of birds led
Darwin to question the idea that species were unchanging.
His collections from the Galápagos Islands suggested that
islands or isolation could give rise to new species.

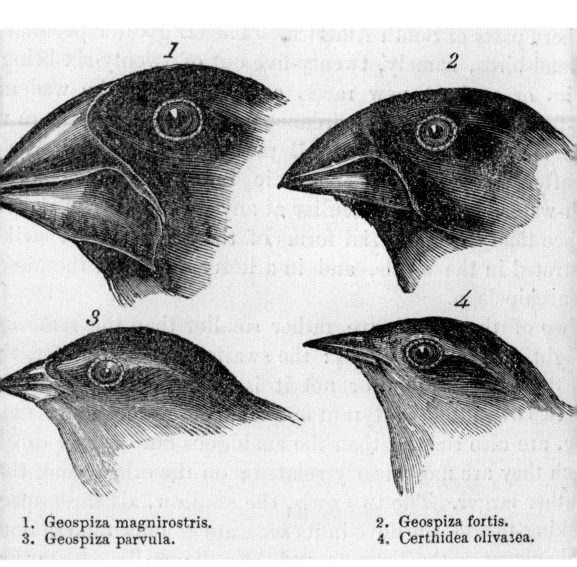

1. Geospiza magnirostris.
2. Geospiza fortis.
3. Geospiza parvula.
4. Certhidea olivasea.

# April 2009

13 Monday                                        Easter Monday (UK & Cana

14 Tuesday

15 Wednesday

16 Thursday                                      Last day of Passover (Jewi

17 Friday

18 Saturday

19 Sunday

1882 Darwin dies at his home, Down
House in Downe, Kent.

**Down House**
Less than two miles from the village of Downe in Kent,
Down House, purchased by Darwin's father for £2000,
became Darwin's home for some forty years.

# April 2009

## 20 Monday

## 21 Tuesday

## 22 Wednesday

## 23 Thursday

St. George's Day (Engla

1851 The death of Darwin's beloved
daughter, Annie, aged 10, dispels in
Darwin any belief in a caring God.

## 24 Friday

## 25 Saturday

Anzac Day (Australia), New Moon

## 26 Sunday

1882 Darwin is buried beneath
the monument to Newton at
Westminster Abbey.

**Darwin plaque**
Shortly after Darwin's death a memorial committee was
formed by the Royal Society in which it was decided to erect
a bronze plaque of Darwin in Westminster Abbey, and
commission a marble statue for the Natural History Museum

27 Monday

28 Tuesday

29 Wednesday

30 Thursday

I Friday

2 Saturday

1808 Birth of Emma Darwin (née
Wedgwood), wife of Charles Robert
Darwin and daughter of Josiah
Wedgwood II.

3 Sunday

**Emma Darwin, 1808-1896**
Emma was raised in a wealthy household and was an
educated, talented and well-travelled young woman
as well as a committed Christian. Emma and Charles
were first cousins and their marriage produced ten children.

# May 2009

4 **Monday**

1825 Birth of Thomas Huxley who became
a passionate defender of Darwin's theory of
evolution through natural selection, and
was known as 'Darwin's bulldog'.

Early May Bank Holiday (U

5 **Tuesday**

Cinco de Mayo (U

6 **Wednesday**

7 **Thursday**

8 **Friday**

9 **Saturday**

Full Moon

10 **Sunday**

Mother's Day (US & Canad

**Thomas Huxley, 1825-1895**
Huxley was an excellent comparative anatomist and taught
biological sciences at the Royal School of Mines.

II   Monday

I2   Tuesday

I3   Wednesday

I4   Thursday

I5   Friday

1862  Publication of Darwin's book *On the various contrivances by which British and foreign orchids are fertilised by insects.*

I6   Saturday

I7   Sunday

**Ophrys apifera**
This bee orchid is one of the many Franz Bauer's paintings of orchids that were used by George Sowerby to help illustrate Darwin's publication of British and foreign orchids.

*Ophrys apifera* Huds.

## 18 Monday

1832 Writing from Rio de Janeiro, Darwin admires Alexander Humboldt, "he alone gives any notion, of the feelings which are raised in the mind on first entering the Tropics."

## 19 Tuesday

## 20 Wednesday

## 21 Thursday

## 22 Friday

1860 Darwin comments about Asa Gray's theistic beliefs, "I cannot see...evidence of design and beneficence on all sides of us. There seems to me too much misery in the world."

## 23 Saturday

## 24 Sunday

New Moon

**Alexander Humboldt at the foot of Mount Chimborazo**
Humboldt's *A Personal Narrative* about his travels in South America was one of the greatest influences on Darwin as a young man. Darwin took it with him on his voyage to South America on HMS *Beagle* and read it several times over.

25 Monday

Spring Bank Holiday (UK), Memorial Day (

26 Tuesday

27 Wednesday

28 Thursday

29 Friday

30 Saturday

31 Sunday

Pentecost (Christ

1876 Darwin starts work on his
autobiography, finishing it by 3 August
the same year.

**Darwin's study, Down House**
Darwin's study was equipped with his scientific instruments
books and a comfortable chair with attached wheels enablin
him to reach most things without having to rise. Close to th
fire was a bed for his dog, Polly.

## 1 Monday

## 2 Tuesday

## 3 Wednesday

1836 HMS *Beagle* reaches Cape Town and Darwin visits Sir John Herschel. They discuss the evolving landscape, and the "mystery of mysteries", the appearance of new species.

## 4 Thursday

## 5 Friday

## 6 Saturday

## 7 Sunday

Trinity Sunday (Christian), Full Moon

**Darwin's finches**
Darwin's birds from the Galápagos Islands were, he believed, a collection of wrens, finches, gross-beaks and blackbirds. In fact, once John Gould had examined them, he found that they were twelve different species of finches.

## 8 Monday

---

## 9 Tuesday

1885 Unveiling of the statue of Darwin
on the Central Hall staircase, Natural
History Museum, London.

---

## 10 Wednesday

---

## 11 Thursday

Corpus Christi (Christi

---

## 12 Friday

---

## 13 Saturday

1817 Birth of Joseph Dalton Hooker, a
botanist, who became one of Darwin's
closest friends and supporter of his
ideas on natural selection.

---

## 14 Sunday

Flag Day ((

---

**Charles Robert Darwin, 1809-1882**
The unveiling of the statue of Charles Darwin in the Natural
History Museum, London, published in *The Graphic* on the
20th June 1885.

TRUSTEES

CHARLES DARWIN

BORN
FEB 12 1809
DIED
APRIL 19 1882

ING THE STATUE OF THE LATE CHARLES DARWIN IN THE NATURAL HISTORY MUSEUM, SOUTH KENSINGTON

I5 Monday

I6 Tuesday

I7 Wednesday

I8 Thursday

1858 Darwin receives from Alfred Russel
Wallace a paper on natural selection titled
*On the Tendency of Varieties to Depart
Indefinitely from the Original Type.*

I9 Friday

2O Saturday

2I Sunday

Father's Day, Summer solstice (UK &

**Alfred Russel Wallace, 1823-1913**
Independently of Darwin, the naturalist and explorer Alfred
Russel Wallace developed the theory of evolution through
natural selection when on the Island of Ternate in Indonesia
in 1858. He immediately wrote to Darwin on the subject.

Alfred R. Wallace

## 22 Monday

New Moo

## 23 Tuesday

## 24 Wednesday

## 25 Thursday

## 26 Friday

## 27 Saturday

**June 1834** HMS *Beagle* reaches the Pacific
Ocean, two and a half years after setting sail.

## 28 Sunday

### *Rhea darwinii*
The lesser rhea as Darwin probably saw it in South America
1833, when he rescued a specimen from the Christmas dinn
table. The ornithologist John Gould identified it as a new
species and named it after its discoverer.

*Rhea Darwinii.*

## 29 Monday

## 30 Tuesday

1860 The first public debate on natural selection at a British Association for the Advancement of Science meeting, between Bishop Wilberforce, Thomas Huxley and Joseph Hooker.

## I Wednesday

Canada Da

1858 Darwin's and Alfred Russel Wallace's papers on evolution are delivered jointly to the Linnean Society, one read by Joseph Hooker, the other by Charles Lyell.

## 2 Thursday

## 3 Friday

## 4 Saturday

Independence Day (U

## 5 Sunday

**Journal of the Proceedings of the Linnean Society, 1858**
Faced with the prospect of someone else publishing before him, Darwin's friends, Lyell and Hooker, arranged for Wallace and Darwin to have papers presented to the Linnean Society. This is Wallace's copy of the published papers.

[*From the* JOURNAL *of the* PROCEEDINGS OE THE LINNEAN SOCIETY *for*
*August* 1858.]

On the Tendency of Species to form Varieties; and on the Per-
petuation of Varieties and Species by Natural Means of
Selection. By CHARLES DARWIN, Esq., F.R.S., F.I.S., &
F.G.S., and ALFRED WALLACE, Esq. Communicated by Sir
CHARLES LYELL, F.R.S., F.L.S., and J. D. HOOKER, Esq.,
M.D., V.P.R.S., F.L.S., &c.

[Read July 1st, 1858.]

London, June 30th, 1858.

MY DEAR SIR,—The accompanying papers, which we have the
honour of communicating to the Linnean Society, and which all
relate to the same subject, viz. the Laws which affect the Pro-
duction of Varieties, Races, and Species, contain the results of the
investigations of two indefatigable naturalists, Mr. Charles Darwin
and Mr. Alfred Wallace.

These gentlemen having, independently and unknown to one
another, conceived the same very ingenious theory to account for
the appearance and perpetuation of varieties and of specific forms
on our planet, may both fairly claim the merit of being original
thinkers in this important line of inquiry; but neither of them
having published his views, though Mr. Darwin has for many
years past been repeatedly urged by us to do so, and both authors
having now unreservedly placed their papers in our hands, we
think it would best promote the interests of science that a selection from them should be laid before the Linnean Society.

Taken in the order of their dates, they consist of :—

1. Extracts from a MS. work on Species*, by Mr. Darwin, which
was sketched in 1839, and copied in 1844, when the copy was read
by Dr. Hooker, and its contents afterwards communicated to Sir
Charles Lyell. The first Part is devoted to "The Variation of
Organic Beings under Domestication and in their Natural State;"
and the second chapter of that Part, from which we propose to
read to the Society the extracts referred to, is headed, " On the
Variation of Organic Beings in a state of Nature; on the Natural
Means of Selection; on the Comparison of Domestic Races and
true Species."

2. An abstract of a private letter addressed to Professor Asa
Gray, of Boston, U.S., in October 1857, by Mr. Darwin, in which

* This MS. work was never intended for publication, and therefore was not
written with care.—C. D. 1858.

# July 2009

6 Monday

7 Tuesday

Full Moon

8 Wednesday

9 Thursday

July 1837 Darwin began his 'B' notebook
in which he put down his thoughts on
the subject of transmutation.

IO Friday

II Saturday

I2 Sunday

**Joseph Dalton Hooker, 1817-1911**
A botanist and, in later life, Director of the Botanic Gardens,
Kew, Joseph Hooker was one of Darwin's closest friends. He
publicly supported Darwin's theory of natural selection when
he published *Introductory Essay to the Flora Tasmaniae* in 1859

# July 2009

13 Monday

14 Tuesday

15 Wednesday

1817 Death of Susannah Darwin, Charles's
mother. In later life Darwin commented that
"I can remember hardly anything about her".

16 Thursday

17 Friday

18 Saturday

19 Sunday

*Carollia brevicauda*, silky short-tailed bat
In studying the skeletal structure of bats Darwin found that the
bone of a bat's wing corresponded with those in a human hand
providing evidence that mammals shared a common ancestry.

*Desmodus D'Orbignyi.*

20 Monday

21 Tuesday

22 Wednesday                                                                    New Moon

23 Thursday

24 Friday

25 Saturday

26 Sunday

1843 Writing to the entomologist George
Waterhouse about transmutation, Darwin states
that the genealogical relationship between species
should be based upon common descent.

**Letter to Emma, 5th July 1844**
This letter contained Darwin's "most solemn & last request"
to his wife Emma, concerning the publication of his evolution
essay, asking her to "devote £400 to its publication" in the
case of his death.

Down July 5ᵗ — 1844    4

My Dear Emma

I have just finished my sketch of my species
theory. If, as I believe, that my theory in time be accepted even by one competent judge, it
will be a considerable step in science.
I therefore write this, in case of my sudden
death, as my most solemn & last request, which
I am sure you will consider the same as if
legally entered in my will, that you will
devote 400£ to its publication & further will
yourself, or through Hensleigh, take trouble in
promoting it. — I wish that my sketch be
given to some competent person, with this
sum to induce him to take trouble in its
improvement & enlargement. — I give to him all
my Books on Natural History, which are either
scored or have references at end to the pages,
begging him carefully to look over & consider
such passages as actually bearing on & possibly
bearing on the subject. I wish you to make a list of all such books as some temptation to a extra...
I also repeat that you...

## 27 Monday

## 28 Tuesday

## 29 Wednesday

## 30 Thursday

1851 Darwin and his family visit the Great
Exhibition at Crystal Palace. Darwin is so
amazed he makes several visits.

## 31 Friday

## I Saturday

## 2 Sunday

**South America**
Frogs and toads from South America including
*Phryniscus nigricans*, found by Charles Darwin in
South America whilst on the voyage of HMS *Beagle*,
which showed a distinct antipathy to water.

Drawn from Nature on Stone by B. Waterhouse Hawkins.

Printed by C. Hullmandel.

1. 2. *Rhinoderma Darwinii.*
3. 4. 5. *Phryniscus nigricans.* } *Nat. size.*
6. *Uperodon ornatum.*

## 3 Monday

**August 1835** To his sister Catherine, Darwin writes "I am very anxious to see the Galapagos Islands...both the Geology & Zoology cannot fail to be very interesting".

## 4 Tuesday

## 5 Wednesday

## 6 Thursday

Full Moon

## 7 Friday

## 8 Saturday

## 9 Sunday

**The Zoology of the Voyage of HMS Beagle, 1841**
The bird descriptions and drawings for this work were by the artist and ornithologist John Gould. There are fifty hand-coloured plates, including nine of Darwin's finches from the Galápagos Archipelago.

## 10 Monday

## 11 Tuesday

1830 Holidays were spent collecting insects.
Darwin sets out to visit Barmouth in Wales
where he amassed a large collection of
beetles and climbed Mount Snowdon.

## 12 Wednesday

## 13 Thursday

## 14 Friday

## 15 Saturday

## 16 Sunday

**Darwin's beetles**
As a child Darwin was an avid collector of natural history,
including birds' eggs, shells and insects, especially beetles.
On the HMS *Beagle* voyage, he amassed a magnificent
collection of beetles from around the world.

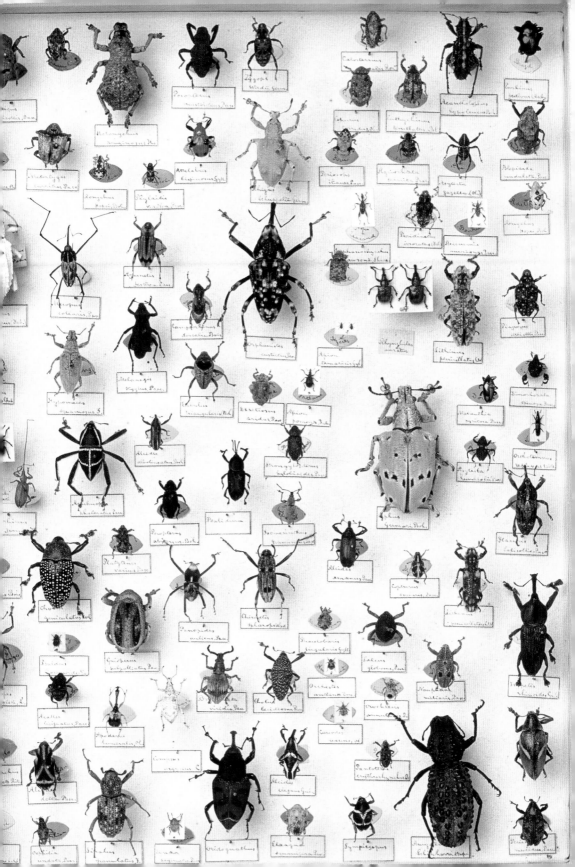

## 17 Monday

## 18 Tuesday

## 19 Wednesday

## 20 Thursday

New Moo

## 21 Friday

## 22 Saturday

Ramadan begins (Isla

## 23 Sunday

**August 1836** Heading home, HMS *Beagle*
makes a detour to South America, producing
despair in Darwin: "I loathe, I abhor the sea,
and all ships which sail on it".

**Pseudocyphellaria freycinettii**
HMS *Beagle* called at Tierra Del Fuego several times
and Darwin was able to make extensive notes on
the local fauna and flora. He collected many specimens
including this lichen which mostly occurs on soil
and rocks in open moor and grassland.

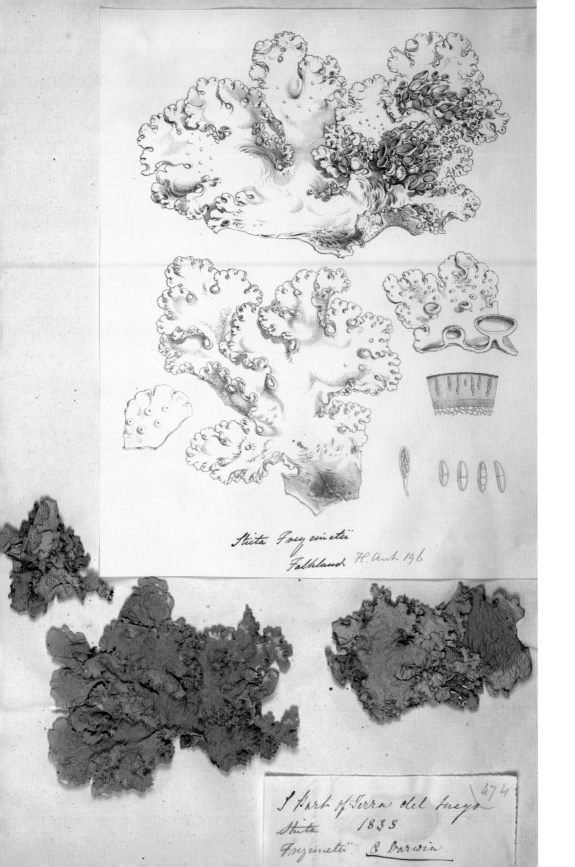

Sticta Freycinetii

Falkland H. Ant. 196

I Part of Terra del Fuego
Sticta 1833
Freycinetii C Darwin

474

## 24 Monday

## 25 Tuesday

## 26 Wednesday

## 27 Thursday

## 28 Friday

## 29 Saturday

1831 A letter from Henslow relates the news
that Darwin has been invited to sail as
Captain FitzRoy's companion on HMS *Beagle*.

## 30 Sunday

**Mylodon darwinii**
Illustration of the jaw bone of an extinct, giant, ground
sloth collected by Darwin in South America. Paleontologist
and comparative anatomist Richard Owen examined
and described the fossil mammals for *The Zoology of
the Voyage of HMS Beagle*.

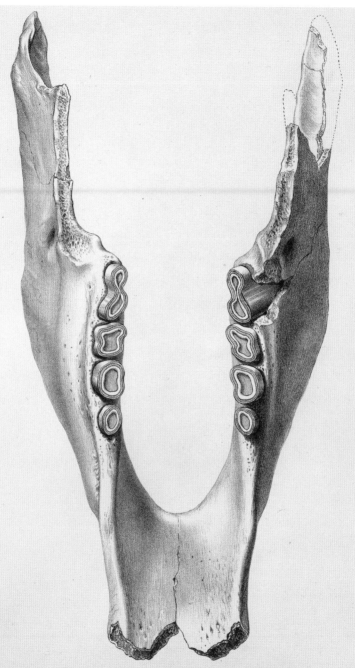

G. Scharf del et lithog.

Printed by C. Hullmandel.

Mylodon. ⅔ Nat. Size.

Published by Smith, Elder & Co. 65 Cornhill, London.

31 Monday

Summer Bank Holiday (U

---

1 Tuesday

---

2 Wednesday

---

3 Thursday

---

4 Friday

Full Moon

---

5 Saturday

1831 *"Gloria in excelsis* is the most moderate beginning I can think of ...", Darwin writes to Henslow after being accepted to join the *Beagle* voyage.

6 Sunday

---

**The Malay Archipelago, 1869**
The two volume work by Alfred Russel Wallace about his travels in Malaysia and Indonesia was dedicated to Charles Darwin. Wallace corresponded with Darwin on a whole range of issues concerning the theory of evolution.

TO

# CHARLES DARWIN,

*AUTHOR OF "THE ORIGIN OF SPECIES,"*

I Dedicate this Book,

NOT ONLY

AS A TOKEN OF PERSONAL ESTEEM AND FRIENDSHIP

BUT ALSO

TO EXPRESS MY DEEP ADMIRATION

FOR

His Genius and his Works.

# September 2009

7 Monday

8 Tuesday

9 Wednesday

10 Thursday

1846 Darwin completes his *Geological Observations on South America*.

11 Friday

12 Saturday

13 Sunday

**Dendroligotrichum dendroides**
Darwin collected this moss on the rugged islands of Tierra del Fuego in 1833. It grows in dense clumps on forest floors in Chile and New Zealand. Darwin, like natural scientists today, saw the value of a specimen as evidence of a species living in a particular place at a particular time.

Polytri. dendroides.
S. part of Terra del Fuego.
1833
C. Darwin 465.

Polytrich. dendroides.
Chiloe    Capt. King

# September 2009

## 14 Monday

## 15 Tuesday

## 16 Wednesday

1835 HMS *Beagle* reaches the Galápagos
Archipelago. Darwin is fascinated by the
adaptation of the marine iguana to
swimming and feeding in the ocean.

## 17 Thursday

1842 Darwin and his family move to
Down House in the village of Downe,
Kent where he lived out the rest of his life.

## 18 Friday

New Moon

## 19 Saturday

Rosh Hashanah (Jewis

## 20 Sunday

Jewish New Ye

**Darwin's greenhouse**
Darwin spent many hours in his greenhouse, growing orchid
and carnivorous plants, observing how these plants were
pollinated by insects and studying cross-fertilization.

# September 2009

## 21 Monday

**1835** On the Galápagos Islands, Darwin meets with giant tortoises describing them in his diary as "old fashioned antediluvian animals".

## 22 Tuesday

Autumnal Equinox – Autumn begins (UK & US)

## 23 Wednesday

## 24 Thursday

## 25 Friday

## 26 Saturday

## 27 Sunday

**Drawing of giant tortoise skeleton, 1986**
Darwin discovered that the tortoise on each Galápagos island was slightly different. He took great delight in riding on the shells of the tortoises but declared that he found it difficult to keep his balance.

## 28 Monday

1838 Darwin begins reading *An essay on the Principle of Population* by Thomas Malthus, a work that inspired Darwin to think seriously about natural selection.

## 29 Tuesday

## 30 Wednesday

## 1 Thursday

## 2 Friday

1836 HMS *Beagle* returns from its voyage, after almost five years, docking at Falmouth.

## 3 Saturday

## 4 Sunday

Full Moo

**Charles Robert Darwin, 1809-1882**
A caricature portrait of Darwin published in the magazine *Vanity Fair* in September 1871.

5 Monday

6 Tuesday

7 Wednesday

8 Thursday

1835 Darwin explores St. James Island
in the Galápagos Archipelago, remaining
there for nine days collecting and observing.

9 Friday

10 Saturday

1840 Darwin visits the Zoology
Department of the British Museum,
London and signs the visitor book.

11 Sunday

**Letter to HG Bronn, 14th February 1860**
Darwin thanks Bronn, who supervised the German translation
of *On the Origin of Species*, and discusses the importance of
"choosing good German terms for natural selection". Here,
is also a rare example of Darwin's full signature.

Lamarckian doctrine (which I reject) of habits of life being all-important.

Man has altered & thus improved the English Race-Horse by selecting successive fleeter individuals; & I believe, owing to the struggle for existence, that similar slight variations in a wild Horse, if advantageous to it, would be selected or preserved by nature: Hence natural Selection.

But I apologise for troubling you with these remarks on the importance of choosing good German terms for "natural selection."

With my heart-felt thanks & with sincere respect

I remain Dear Sir

yours very sincerely

Charles Darwin.

I am very much obliged for your "Stuffengang &c", which I am now reading: I wish I knew what was the authority for a Batrachian in the New Hebrides.

---

I2 Monday                                              Columbus Day (US), Thanksgiving (Cana

---

I3 Tuesday

---

I4 Wednesday

---

I5 Thursday

**1827** Darwin is accepted into Christ's
College, Cambridge, to study for the
clergy, and graduates in 1831.

---

I6 Friday

---

I7 Saturday

---

I8 Sunday                                                                    New Moon

---

**Angraecum sesquipedale**
Nectar of the comet orchid lies at the bottom of a 30 cm
(12 in) long tube. Darwin predicted that the pollinator must
have an equally long tongue. Twenty-one years later the
insect responsible, the hawkmoth, was discovered.

## 19 Monday

## 20 Tuesday

## 21 Wednesday

## 22 Thursday

**October 1825** At the age of 16, Darwin goes to Edinburgh University to study medicine, but does not enjoy it and leaves in April 1827 without a degree.

## 23 Friday

## 24 Saturday

United Nations [

## 25 Sunday

*Lutreolina crassicaudata*, **thick-tailed opossum**
A specimen of this opossum was caught by Darwin in Maldonad
Uruguay. He wrote that his excursions to Maldonado over a peri
of three years resulted in him collecting several quadrupeds, 80
kinds of birds, and many reptiles, including nine species of snak

*Didelphis crassicaudata*

# October › November 2009

**26** Monday

**27** Tuesday

**October 1832** Darwin receives Lyell's second
volume of *of Principles of Geology*. The work
denounces Lamarck's ideas on evolution, providing
Darwin with some interesting ideas to dwell on.

**28** Wednesday

**29** Thursday

**30** Friday

**31** Saturday

Halloween

**1** Sunday

All Saints Day, Daylight Saving Time ends (UK & US)

**Shu No Kigen, Tokyo, 1914**
This is a first edition of the Japanese translation of *The Origin
of Species*. The book was published as a pocket-size edition in
five volumes and includes a folding image of a speciation tree

# November 2009

**2**  Monday
Full Moon

**3**  Tuesday
Election Day (U

**4**  Wednesday

**5**  Thursday
Bonfire Night (U

**6**  Friday
1826 Darwin enters his second year of
medical school. Bored, he spends much of his
time with William MacGillivray, the curator of
Robert Jameson's Natural History Museum.

**7**  Saturday

**8**  Sunday

*Pinguinus impennis,* **great auk**
The naturalist William MacGillivray was also a gifted artist
who produced stunning bird portraits. He possessed a wealt
of knowledge concerning British natural history and was no
doubt an inspiration to Darwin.

9 Monday

10 Tuesday

11 Wednesday                                    Veterans Day (US), Remembrance Day (Canad

12 Thursday

13 Friday

14 Saturday

1797 Birth of Charles Lyell, author of
*Principles of Geology*. Lyell was
instrumental in getting Darwin to
publish *On the Origin of Species*.

15 Sunday

**Charles Lyell, 1797-1875**
A Scottish lawyer and geologist, Lyell was one of the most
influential scientists of his day. Darwin read his *Principles
of Geology* on the *Beagle* voyage and, when he returned
to England, they became close personal friends.

16 Monday

New Moon

17 Tuesday

**November 1833** Returning from an expedition to the Mercedes region in Uruguay, Darwin discovers a fossil head of a hippo-like animal, *Toxodon*.

18 Wednesday

19 Thursday

20 Friday

**November 1846** Darwin starts his research on barnacles, a study that consumes much of his time for the next eight years.

21 Saturday

22 Sunday

**Pollicipes mitella**
Barnacles such as *Pollicipes* were fascinating to Darwin, especially their reproductive biology. As a self-taught zoologist, he spent eight years of his life dedicated to working on the taxonomy of barnacles.

mitella, Linn.

11

---

**23** Monday

---

**24** Tuesday

**1859**  *On the Origin of Species by Means of Natural Selection, or the Preservation of Favoured Races in the Struggle for Life is published.* 1250 copies are printed.

---

**25** Wednesday

---

**26** Thursday

Thanksgiving Day (

---

**27** Friday

---

**28** Saturday

**November 1872**  Publication of *The Expression of the Emotions in Man and Animals,* a work that continued his study of human origin started with *The Descent of Man.*

---

**29** Sunday

---

**Title page of *On the Origin of Species*, 1859**
Based on the essay written in 1844, and substantially expanded in the 15 months prior to publication, the work changed the way we view and understand the natural world and mankind's place within it.

19 JUN 1926

ON

# THE ORIGIN OF SPECIES

## BY MEANS OF NATURAL SELECTION,

OR THE

PRESERVATION OF FAVOURED RACES IN THE STRUGGLE
FOR LIFE.

By CHARLES DARWIN, M.A.,

FELLOW OF THE ROYAL, GEOLOGICAL, LINNÆAN, ETC., SOCIETIES;
AUTHOR OF 'JOURNAL OF RESEARCHES DURING H. M. S. BEAGLE'S VOYAGE
ROUND THE WORLD.'

LONDON:
JOHN MURRAY, ALBEMARLE STREET.
1859.

*The right of Translation is reserved.*

## 30 Monday

## 1 Tuesday

**December 1831** Preparation for the voyage is well underway and Darwin purchases supplies: two pistols, a rifle and scientific instruments including an inclinometer.

## 2 Wednesday

Full Moon

## 3 Thursday

## 4 Friday

## 5 Saturday

## 6 Sunday

**Early December 1838** Darwin starts his 'E' notebook, the fourth and last of his notes on transmutation, listing the three principles that will account for natural selection.

**Original manuscript**
One of five draft manuscript pages "On Instinct", part of *On the Origin of Species*. It appears in chapter seven of the published work.

[Ch. 7. Instinct]

p. 258 middle. See 5th edn 1869. 237

As to bodily organs. Changes of instinct may sometimes be facilitated by the same species having different instincts at different period of life, or time of the year, or when placed under different circumstances &c; in which case either one instinct or the other might be preserved by natural selection: and such cases of diversity of instinct in the same species can be shown to occur in nature.]

[Again, as in the case of corporeal structure, & conformably with our theory, the instinct of each species is good for itself, but has never, as far as we can judge, been produced for the exclusive good of others.

p. 259

Each species tries to take advantage of the instincts of others. In some cases certain instincts cannot be considered as absolutely perfect; but as details on this and other such heads are not indispensable, they shall not be here given.

p. 259 near bottom

p. 260

As some degree of variation in instincts under a state of nature, & the inheritance of such variations, is the indispensable foundation for natural selection to act on; it would be highly desirable here to give instances of variation; but want of space prevents me. I can only say that variations certainly do occur, for instance in the migratory instincts of

near top of p. 260

7   Monday

8   Tuesday

9   Wednesday

IO   Thursday

II   Friday

I2   Saturday                       Festival of Lights, Hanukkah (Jewis

**1731**   Birth of Erasmus Darwin,
grandfather of Charles, and exponent of
the theory of evolution in his book
*Zoonomia, or, The Laws of Organic Life.*

I3   Sunday

**Erasmus Darwin, 1731-1802**
Charles Darwin's grandfather, Erasmus, was a successful
physician, inventor, naturalist, poet and was fervently
anti-slavery. His writings anticipated many revolutionary
ideas about the universe and the natural world, including
evolution and the Big Bang theory.

14 Monday

15 Tuesday

16 Wednesday                                                    New Moon

17 Thursday

18 Friday

1832 HMS *Beagle* makes its first visit to
Tierra del Fuego and a party of men,
including Darwin, is sent to communicate
with the indigenous people, the Fuegians.

19 Saturday

20 Sunday

**Fuegians**
The first encounter with the Fuegians left a distinct impressio
on Darwin: "It was without exception the most curious and
interesting spectacle I ever beheld", wrote Darwin in his
*Journal of Researches*.

HEAD OF FUEGIAN.          HEAD OF A TAHITIAN.

FUEGIAN VILLAGE.

# December 2009

21 Monday

Winter Solsti

22 Tuesday

23 Wednesday

24 Thursday

25 Friday

Christmas Day (Christia

26 Saturday

Boxing Day (UK & Canada), Australia Day (Australi

27 Sunday

1831 HMS *Beagle* sets sail leaving
England for its voyage around the world.

**The *Beagle* in the Strait of Magellan**
The mission of the *Beagle* voyage was to chart the waters
of South America, particularly the treacherous waters of
the Strait of Magellan, on the continent's southern tip.

---

28 Monday

Bank Holiday (U

---

29 Tuesday

---

30 Wednesday

**Late December 1859** The Prime Minister suggests
to Queen Victoria that Darwin is given a knighthood,
but Bishop Wilberforce intervenes; such honours
would imply support for Darwin's theory.

---

31 Thursday

New Year's Eve, Full Moon

---

1 Friday

New Year's D

**1839** Darwin moves to a house in
London with a 30-metre garden, and
starts a routine of walking up and down
his garden on a daily basis.

---

2 Saturday

---

3 Sunday

---

**Sandwalk at Down House**
Down House was more than a home for Darwin, it was
his laboratory, place of research and a retreat from the
hectic outside world. The gardens included Darwin's
thinking path, called the Sandwalk.

|     | JAN | FEB | MAR | APR | MAY | JUNE | JULY | AUG | SEPT | OCT | NOV | DEC |
| --- | --- | --- | --- | --- | --- | --- | --- | --- | --- | --- | --- | --- |
| M   |     | 1   | 1   |     |     |      |      |     |      |     | 1   |     |
| T   |     | 2   | 2   |     |     | 1    |      |     |      |     | 2   |     |
| W   |     | 3   | 3   |     |     | 2    |      |     | 1    |     | 3   | 1   |
| Th  |     | 4   | 4   | 1   |     | 3    | 1    |     | 2    |     | 4   | 2   |
| F   | 1   | 5   | 5   | 2   |     | 4    | 2    |     | 3    | 1   | 5   | 3   |
| S   | 2   | 6   | 6   | 3   | 1   | 5    | 3    |     | 4    | 2   | 6   | 4   |
| S   | 3   | 7   | 7   | 4   | 2   | 6    | 4    | 1   | 5    | 3   | 7   | 5   |
| M   | 4   | 8   | 8   | 5   | 3   | 7    | 5    | 2   | 6    | 4   | 8   | 6   |
| T   | 5   | 9   | 9   | 6   | 4   | 8    | 6    | 3   | 7    | 5   | 9   | 7   |
| W   | 6   | 10  | 10  | 7   | 5   | 9    | 7    | 4   | 8    | 6   | 10  | 8   |
| Th  | 7   | 11  | 11  | 8   | 6   | 10   | 8    | 5   | 9    | 7   | 11  | 9   |
| F   | 8   | 12  | 12  | 9   | 7   | 11   | 9    | 6   | 10   | 8   | 12  | 10  |
| S   | 9   | 13  | 13  | 10  | 8   | 12   | 10   | 7   | 11   | 9   | 13  | 11  |
| S   | 10  | 14  | 14  | 11  | 9   | 13   | 11   | 8   | 12   | 10  | 14  | 12  |
| M   | 11  | 15  | 15  | 12  | 10  | 14   | 12   | 9   | 13   | 11  | 15  | 13  |
| T   | 12  | 16  | 16  | 13  | 11  | 15   | 13   | 10  | 14   | 12  | 16  | 14  |
| W   | 13  | 17  | 17  | 14  | 12  | 16   | 14   | 11  | 15   | 13  | 17  | 15  |
| Th  | 14  | 18  | 18  | 15  | 13  | 17   | 15   | 12  | 16   | 14  | 18  | 16  |
| F   | 15  | 19  | 19  | 16  | 14  | 18   | 16   | 13  | 17   | 15  | 19  | 17  |
| S   | 16  | 20  | 20  | 17  | 15  | 19   | 17   | 14  | 18   | 16  | 20  | 18  |
| S   | 17  | 21  | 21  | 18  | 16  | 20   | 18   | 15  | 19   | 17  | 21  | 19  |
| M   | 18  | 22  | 22  | 19  | 17  | 21   | 19   | 16  | 20   | 18  | 22  | 20  |
| T   | 19  | 23  | 23  | 20  | 18  | 22   | 20   | 17  | 21   | 19  | 23  | 21  |
| W   | 20  | 24  | 24  | 21  | 19  | 23   | 21   | 18  | 22   | 20  | 24  | 22  |
| Th  | 21  | 25  | 25  | 22  | 20  | 24   | 22   | 19  | 23   | 21  | 25  | 23  |
| F   | 22  | 26  | 26  | 23  | 21  | 25   | 23   | 20  | 24   | 22  | 26  | 24  |
| S   | 23  | 27  | 27  | 24  | 22  | 26   | 24   | 21  | 25   | 23  | 27  | 25  |
| S   | 24  | 28  | 28  | 25  | 23  | 27   | 25   | 22  | 26   | 24  | 28  | 26  |
| M   | 25  |     | 29  | 26  | 24  | 28   | 26   | 23  | 27   | 25  | 29  | 27  |
| T   | 26  |     | 30  | 27  | 25  | 29   | 27   | 24  | 28   | 26  | 30  | 28  |
| W   | 27  |     | 31  | 28  | 26  | 30   | 28   | 25  | 29   | 27  |     | 29  |
| Th  | 28  |     |     | 29  | 27  |      | 29   | 26  | 30   | 28  |     | 30  |
| F   | 29  |     |     | 30  | 28  |      | 30   | 27  |      | 29  |     | 31  |
| S   | 30  |     |     |     | 29  |      | 31   | 28  |      | 30  |     |     |
| S   | 31  |     |     |     | 30  |      |      | 29  |      | 31  |     |     |
| M   |     |     |     |     | 31  |      |      | 30  |      |     |     |     |
| T   |     |     |     |     |     |      |      | 31  |      |     |     |     |

Library of Congress Cataloging-in-Publication Data Available

2    4    6    8    10    9    7    5    3    1

Published in 2008 by Sterling Publishing Co., Inc.
387 Park Avenue South, New York, NY 10016

First published in Great Britain in 2008 by the Natural History Museum
Cromwell Road, London SW7 5BD

Distributed in Canada by Sterling Publishing
c/o Canadian Manda Group. 165 Dufferin Street
Toronto, Ontario, Canada M6K 3H6

For information about custom editions, special sales, premium and
corporate purchases, please contact Sterling Special Sales
Department at 800-805-5489 or specialsales@sterlingpub.com.

Manufactured in China

Sterling ISBN: 978-1-4027-6244-4

Edited by Judith Magee
Designed by Studio Gossett
Reproduction by Saxon Photolitho, UK

PICTURE CREDITS:
Week 5 © Angelo Hornak
Week 6 © Whipple Museum of the History of Science, University of Cambridge
Week 10 © Roy Williams and Genesis Publications
Weeks 16, 22, 38 © English Heritage Photo Library
Week 17 © Dean and Chapter of Westminster
Week 18 © The Charles Darwin Trust
2010, Week 1 © Karen James
All other images © Natural History Museum, London

THE AMERICAN MUSEUM OF NATURAL HISTORY in New York City is one
of the world's preeminent scientific, educational and cultural institutions whose
mission is to explore and interpret human cultures and the natural
world through scientific research, education, and exhibitions.
Visit www.amnh.org.